Physical Geography
Laboratory Manual
Eighth Edition

Rainer R. Erhart
Department of Geography
Western Michigan University

Dean G. Butzow
Department of Math and Science
Lincoln Land Community College

Editorial Assistance:
Lisa Butzow

Cover image © 2014, Shutterstock, Inc.

All photos were taken by Dean Butzow unless otherwise noted. All figures created by Dean Butzow.

www.kendallhunt.com
Send all inquiries to:
4050 Westmark Drive
Dubuque, IA 52004-1840

Copyright © 1986, 1993 by Rainer R. Erhart

Copyright © 2000 by Rainer R. Erhart and Rolland N. Fraser

Copyright © 2004, 2014, 2019 by Rainer R. Erhart and Dean G. Butzow

ISBN 978-1-5249-8208-9

Kendall Hunt Publishing Company has the exclusive rights to reproduce this work,
to prepare derivative works from this work, to publicly distribute this work,
to publicly perform this work and to publicly display this work.

All rights reserved. No part of this publication may be reproduced,
stored in a retrieval system, or transmitted, in any form or by any
means, electronic, mechanical, photocopying, recording, or otherwise,
without the prior written permission of the copyright owner.

Published in the United States of America

Contents

Preface...V

UNIT 1   EARTH-SUN RELATIONSHIPS 1

UNIT 2   ATMOSPHERIC TEMPERATURE 11

UNIT 3   ATMOSPHERIC PRESSURE AND WINDS..23

UNIT 4   ATMOSPHERIC MOISTURE AND PRECIPITATION 33

UNIT 5   MID-LATITUDE AND TROPICAL WEATHER 45

UNIT 6   MAPS AND MAP INTERPRETATION...55

UNIT 7   THE EARTH'S CRUST AND EARTH MATERIALS 63

UNIT 8   THE EARTH STRUCTURE AND TECTONIC PROCESSES 69

UNIT 9   WEATHERING AND MASS WASTING...77

UNIT 10   SURFACE AND GROUNDWATER HYDROLOGY 81

UNIT 11   LANDFORMS MADE BY STREAMS...85

UNIT 12   FLUVIAL AND EOLIAN LANDFORMS OF DESERTS.......................89

UNIT 13   GLACIAL LANDSCAPES 93

APPENDIX 1   TABLE OF PHYSICAL EQUIVALENTS

Preface

This <u>Physical Geography Laboratory Manual</u> is designed for a one semester course. Each unit roughly corresponds to one chapter in most of today's textbooks available on the topic.

Individual units follow the traditional sequence of most physical geography textbooks. However, each stands by itself, and changing the sequence as needed should present little difficulty for the instructor. Most units can be completed within a 2-hour laboratory period. Each is presented as a framework for students, which challenges students to rely on creativity and self-discovery, using available resources. We assume that engaging discussions between students and instructor are encouraged during the lab, so that students do not leave the session confused or with incorrect answers.

We have compiled this manual for a student population of mostly non-science majors, who require core science curriculum, but may otherwise never step into another science classroom. Some of the unit questions and many of the supplemental units provide exposure to linkages between these lecture and laboratory ideas with their current or possible future life experiences. Most quantitative work involves English units, although we provide opportunities, at the instructor's prerogative, to convert to and work in metric equivalents. We have added Google Earth exercises within the manual which supplement the lecture and laboratory material. The Google Earth exercises will give the students experience working with Geospatial technologies and deepen their understanding of our physical geography world.

Earth-Sun Relationships   Unit 1

Section I   The Earth's Dimensions

1. What general mathematical name is given to the Earth as a result of having uneven lengths of polar and equatorial diameters?
   _____

2. Why is the Earth in the shape of ellipse?

3. Figure 1.1 shows the polar and equatorial diameters as well as the polar and equatorial circumferences of the Earth. Label the figure with the following dimensions.

   A. Polar diameter _____ miles

   B. Polar circumference _____ miles

   C. Equatorial diameter _____ miles

   D. Equatorial circumference _____ miles

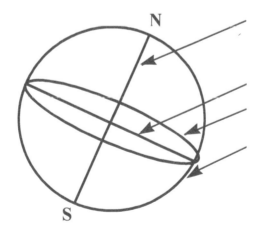

Figure 1.1
Polar and Equatorial Diameters and Circumferences

Section II Measuring Latitude and Longitude Location

Latitude and longitude are abbreviated as shown in the example below. You must be careful to indicate whether a position is located North (N) or South (S) of the equator, and East (E) or West (W) of the Prime Meridian.

Examples: Lat. 9° N, Long. 115° E
Lat. 24 ° 00 ' S, Long. 54° 21 ' W

> **Note:** When minutes are used in the notation of either latitude or longitude, they must also be used in the notation of the other; this is to insure that both are reported with the same spatial precision to be technically correct.

> **Note:** On some maps greater detail is required and hence each degree of latitude or longitude is subdivided into 60 minutes (60') and each minute is subdivided into 60 seconds (60").

Following are two geographic grids, each one representing a simple map. On these maps, North is assumed to be toward the top.

1. On the following figures, determine the latitude and longitude (in degrees and minutes) of the points shown on the conical projection (Figure 1.2) and the cylindrical projection (Figure 1.3).

    A. Point X _____

    B. Point Y _____

    C. Point Z _____

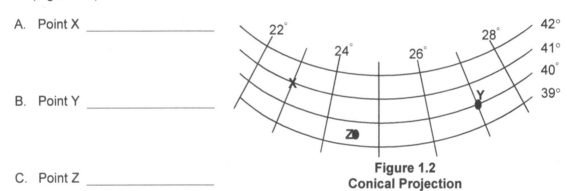

**Figure 1.2
Conical Projection**

2. Different projections produce different grids. Use the rectangular grid below to determine latitude and longitude.

   A. Point P _____

   B. Point Q _____

   C. Point R _____

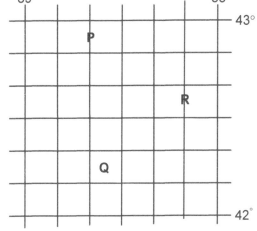

Figure 1.3
Cylindrical Projection

3. Using an atlas, determine the large city located at each of the following points (suggestion: determine the general area in which the city is located by referring to a world map, then find a more detailed regional map to locate the city).

   A. Lat. 42° 22 ' N Long. 83° 10 ' W _____

   B. Lat. 30° 00 ' N Long. 90° 05 ' W _____

   C. Lat. 22° 50 ' S Long. 43° 20 ' W _____

   D. Lat. 39° 55 ' N Long. 116° 23 ' E _____

3

Section III Earth's Orbit

1. Calculate the linear speed of Earth's rotation at the equator.

   _____ mph

2. Calculate the linear speed of Earth's rotation at the latitude of your campus. (For example, Springfield, Illinois at Latitude 40° N has a circumference of 19,029 miles.
   _____ mph.

3. Define the **Angular Rate of Rotation**. Is the Angular Rate of Rotation the same for all latitudes?

4. Define **the Linear Rate of Rotation**. Is the Linear Rate of Rotation the same for all latitudes?

5. An exaggerated diagram of the elliptical orbit of Earth is shown in Figure 1.4.

   A. Give the approximate Earth-Sun distance for:

   July: _____

   January: _____

   B. Indicate the approximate dates and names when Earth is closest and farthest from the sun.

   Closest _____

   Farthest _____

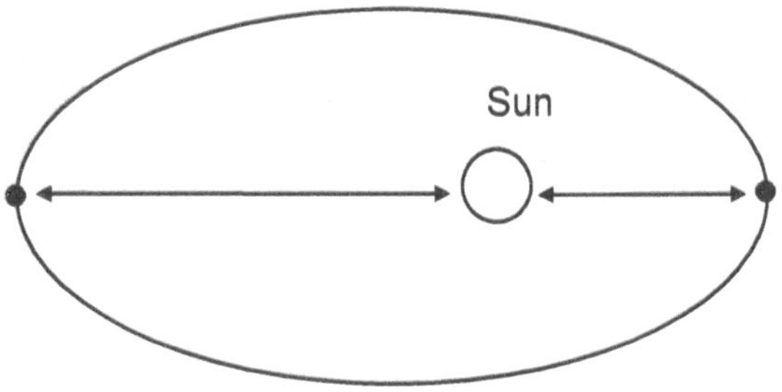

   **Figure 1.4
   Elliptical Orbit**

   C. Indicate (with an arrow) the direction in which Earth revolves around the sun.

   D. What do you deduce from the fact that the sun is closest to Earth during a season in which the temperatures are the lowest in the northern hemisphere?

   E. Earth's rotation is from _____ to _____ (directions) and one full rotation is completed in approximately _____ hours. How many degrees of angular distance does Earth rotate in one hour? _____ In one minute? _____

   F. At a place 30° east of where you are, the sun will rise (sooner, later) by _____ hour(s).

4

Section IV Inclination and Parallelism

Using Figure 1.5 below, study the inclination and parallelism of the Earth's axis. Inclination, along with parallelism and revolution, is responsible for the changing seasons as well as the length of day and night.

1. Draw in the equator and label the season for each of the Earth positions below.

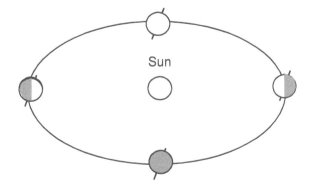

**Figure 1.5
Seasons**

2. What is the approximate start date and the sun's declination for each season listed below?

	Start Date	**Sun's Declination**
A. Vernal Equinox	_____	_____
B. Autumnal Equinox	_____	_____
C. Summer Solstice	_____	_____
D. Winter Solstice	_____	_____

3. What does **parallelism** refer to?

4. What is the latitude and significance of the Tropic of Cancer (Northern Hemisphere perspective)?

5. What is the latitude and significance of the Tropic of Capricorn (Northern Hemisphere perspective)?

6. What is the latitude and significance of the Arctic Circle (Northern Hemisphere perspective)?

7. What is the latitude and significance of the Antarctic Circle (Northern Hemisphere perspective)?

8. If the Earth's axis were not tilted, that is, if it were perpendicular to the plane of the ecliptic:

　A. What would happen to the Earth's seasons?

　B. What would happen to the length of days and nights?

9. Your are planning a trip to Antarctica to see the Emperor penguins. What would be the best month to plan your visit? Why?

Section V   Solstices, Equinoxes and Solar Declination

1. Figure 1.6 shows the summer and winter solstice positions. On this diagram, draw in and label the following:

   A. Circle of illumination (lightly shade the unlighted portion of Earth)

   B. The inclination of Earth's axis (tilt of Earth from vertical)

   C. Arctic and Antarctic Circles and their angular degrees from the equator

   D. Tropics of Cancer and Capricorn and their angular degrees from the equator

   E. The name of the season for each position.

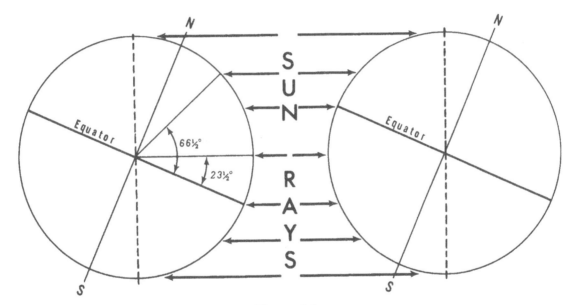

Figure 1.6

Section VI   World Time Zones

Refer to the map of the World Time Zones (Figure 1.7).

1. Name the four standard time zones within the continental United States.

   A. _____   B. _____

   C. _____   D. _____

2. The Rose Bowl game starts at 1:00 pm Pacific Standard Time. The game is televised in Detroit at _____ Eastern Standard Time.

3. A west-bound jet leaves Frankfurt, Germany at 3:00 pm (Germany (Long.15° E) is 1 time zone east of Greenwich, England). Its flight time to New York (Long.75° W) is 7 hours, and its arrival time in New York is _____ Eastern Standard Time.

4. It is noon at Greenwich; your local time is 5:00 pm. What is your longitude?

5. When it is 3:00 am on August 10 at 90° W, what is the time and date at ...

   A. 150° W? _____   B. 150° E? _____

6. When it is 3:00 pm on March 10 at 175° E, what is the time and date at 175° W?

7. The President addressed the nation at 7:30 p.m. Central Time. What time was the address televised?

   A. Mountain Time:_____   B. Eastern Time: _____

   C. Pacific Time:_____   D. Hawaii Time: _____

8. It's 1 p.m. at 15° W Longitude. What is the time at 120° W Longitude?

9. It's 5 p.m. at the Prime Meridian. It is 11 p.m. local time. What is your longitude?

10. An eastbound jet leaves San Diego, CA (120° W Longitude) at 1 p.m. Pacific Time on May 10. It's flight time to Greenwich, England is 7 hours. Its arrival time and day to Greenwich:

11. When it is 9 a.m. on October 15 at 165° W, what is the time and date at 165° E?

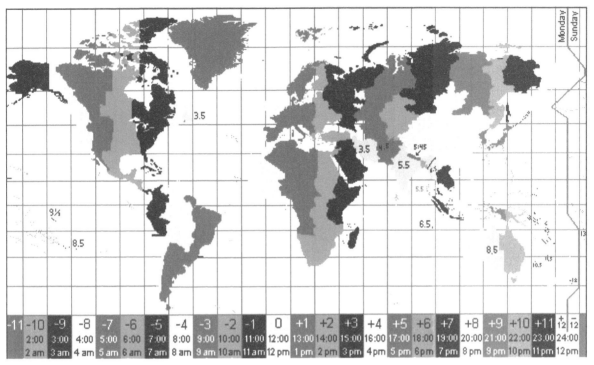

**Figure 1.7
World Time Zones**

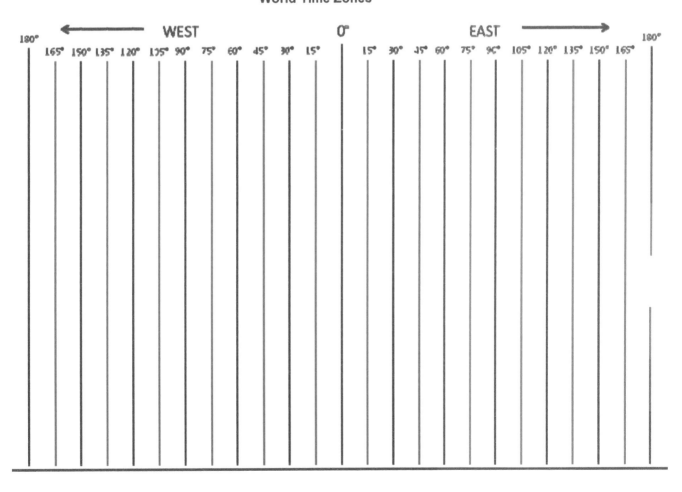

Atmospheric Temperature   Unit 2

Section I   Weather and Climate

1. Define Weather:

2. Define Climate:

Section II   Composition of the Atmosphere

1. Almost 99 percent of the atmosphere is made up of two gases. List these two gases.

   _____

   _____

2. What important functions are performed by the following atmospheric gases and particles?

   A. Ozone   _____

   B. Carbon Dioxide _____

   C. Water Vapor   _____

   D. Dust, Salt   _____

Section III Temperature Controls

1. How do the following global factors control temperature distribution on Earth's surface?

   A. Altitude (include the average rate of temperature decrease)

   B. Latitude

   C. Land-Water Contrast

   D. Ocean Currents (use the Atlantic and Pacific ocean currents that influence the U.S. as examples)

   E. Human Activities

2. Where would you expect to find the largest annual temperature range? Why? Give an example.

3. What is the significance of the troposphere?

   To what approximate elevation above mean sea level does this layer extend? _____ miles

4. Which two factors establish a global heat balance between polar and equatorial latitudes?

Section IV Solar Radiation and Energy Relationships

1. What is the name for the amount of solar radiation received on average at the outermost limit of Earth's atmosphere? _____ _____ What is the percentage of this amount? _____.

2. Approximately what proportion (percentage) of solar radiation (the amount in Question 1) reaches Earth's surface? _____ %

3. List three primary factors preventing much of that solar radiation from reaching the ground.

    A. _____

    B. _____

    C. _____

4. Define albedo.

5. The albedo of fresh fallen snow is between 75%-95%. What does this mean?

6. The albedo of a grassy field is between 10%-30%. What does this mean?

7. Given that the areas sketched in Figure 2.1 are of identical material, but having different albedo, rank them in terms of temperature (afternoon); highest being 1.

Figure 2.1

8. Radiation from the sun is _____ waves while radiation emitted from the earth is _____ waves.

9. What happens to the temperature of a surface:

   A. If the albedo of the surface is very high?

   B. If the albedo of the surface is very low?

10. Describe daytime heating of air inside your car, then relate this phenomenon to the heating of the earth's atmosphere. Use the concept of short waves and long waves in your example.

11. What is the relative atmospheric heat loss (large amount or small amount):

    A. During a summer night in the Arizona desert? Why?

    B. During a cloudy, humid summer night in Kansas? Why?

12. You are watching the weather and the meteorologist says "Tonight it will be cloudy and the clouds will act like a blanket." What does the meteorologist mean by this statement?

13. The weather forecast for **Springfield, Illinois:**
    High Temperature: 80°F   Low Temperature: 72°F

    The weather forecast for **Omaha, Nebraska:**
    High Temperature: 80°F   Low Temperature: 50°F

    Given the above information, which city do you think would report clear skies and which city would report cloudy skies? Why?

14. What time of day do we typically experience our daily maximum and minimum temperatures? Why?

Section V   Temperature Scales

Refer to the Conversion Formulas below to calculate the following temperature conversions.

°C = (°F - 32) ÷ 1.8	°F = (°C X 1.8) + 32
Example:	Example:
68°F = ? °C	10°C = ? °F
= (68-32)	= (10X1.8) + 32
= (36)÷1.8	= 18 + 32
68°F = 20°C	10°C = 50°F

Figure 2.2.
Conversion Formulas

1. Convert 18°C into °F   _____ °F

2. Convert 71.6°F into °C   _____ °C

3. Freezing point of water: _____ °C;   _____ °F

4. Boiling point of water   _____ °C;   _____ °F

5. January temperatures of -40°C have been observed in Northern Minnesota. What is the equivalent temperature in °F?   _____

Section VI Atmospheric Temperature

1. How does the National Weather Service calculate the mean daily temperature?

2. How does the National Weather Service calculate the mean monthly temperature?

3. The mean daily or monthly temperatures can be very misleading. Why?

4. Table 2.1 represents climate data for Springfield, Illinois.
   A. Calculate the mean monthly temperature for each month and fill the data in the table.
   B. Calculate the average daily maximum temperature for the year and fill the in the table.
   C. Calculate the average daily minimum temperature for the year and fill the in the table.
   D. Calculate the average yearly temperature and fill the data in the table.

Month	MEANS °F Daily Maximum	Daily Minimum	Monthly	EXTREMES °F Record Highest	Year	Record Lowest	Year
Jan	34.8	18.7	26.75	73	1909	-22	1884
Feb	39.9	22.6	31.25	78	1930	-24	1905
Mar	52.1	32.2	42.15	91	1907	-12	1960
Apr	64.6	42.4	53.5	90	1986	17	1920
May	74.8	52.6	63.7	101	1934	28	1966
June	83.1	61.9	72.5	104	1934	39	2003
July	86.2	65.4	75.8	112	1954	48	2013
Aug	84.9	63.6	74.25	108	1934	43	1986
Sept	78.9	54.6	66.75	102	2011	31	1899
Oct	66.4	43.8	55.1	93	2006	13	1925
Nov	52.3	33.9	43.1	83	1950	-3	1964
Dec	38.3	22.5	30.4	74	2012	-21	1989
Year	63.03	42.85	52.94				

Table 2.1 Temperature Data for Springfield, Illinois Latitude 39.78°N
(Data from NWS, July 1879-December 2013)

5. Use Table 2.1 and any necessary conversions to complete the following.

   A. Record highest temperature   Yr._____   _____°F   _____°C

   B. Record lowest temperature   Yr._____   _____°F   _____°C

   C. Mean monthly temperature for March   _____°F   _____°C

   D. Mean daily maximum temperature in March   _____°F   _____°C

   E. Mean daily minimum temperature in March   _____°F   _____°C

   F. What is the most recent year to experience a record high temperature? _____ _____°

   G. What is the most recent year to experience a record low temperature? _____ _____°

   H. What is the range in temperature from the record high temperature to the record low temperature? _____

   I. What is the annual range in monthly temperature for Springfield? _____

   J. What is Springfield's highest monthly temperature? What month does it occur? Why?

   K. What is Springfield's lowest monthly temperature? What month does it occur? Why?

6. Fort Bragg, California (Latitude 39.45°N - which is about the same latitude as Springfield, IL) experiences their highest average monthly temperature in July at 57°F. Fort Bragg's lowest average monthly temperature is in January at 46°F. What is the annual average range in temperature for Fort Bragg? _____

7. Using the data from Table 2.1, does Springfield, IL or Fort Bragg, CA experience a larger range in annual temperature? _____ Why do you think this? (You may need to look at a map to locate Fort Bragg's location)

Section VII   Wind Chill and Heat Index

1. Use Table 2.2 to determine the Wind Chill temperature, given the following conditions:

   A. Actual temperature 10° F, Wind Speed 25 miles per hour _____ °F

   B. Actual temperature 40° F, Wind Speed 40 miles per hour _____ °F

   C. Actual temperature 0° F, Wind Speed 40 miles per hour _____ °F

	Actual Air Temperature °F																
Calm	40	35	30	25	20	15	10	5	0	-5	-10	-15	-20	-25	-30	-35	-40
5	36	31	25	19	13	7	1	-5	-11	-16	-22	-28	-34	-40	-46	-52	-57
10	34	27	21	15	9	3	-4	-10	-16	-22	-28	-35	-41	-47	-53	-59	-66
15	32	25	19	13	6	0	-7	-13	-19	-26	-32	-39	-45	-51	-58	-64	-71
20	30	24	17	11	4	-2	-9	-15	-22	-29	-35	-42	-48	-55	-61	-68	-74
25	29	23	16	9	3	-4	-11	-17	-24	-31	-37	-44	-51	-58	-64	-71	-78
30	28	22	15	8	1	-5	-12	-19	-26	-33	-39	-46	-53	-60	-67	-73	-80
35	28	21	14	7	0	-7	-14	-21	-27	-34	-41	-48	-55	-62	-69	-76	-82
40	27	20	13	6	-1	-8	-15	-22	-29	-36	-43	-50	-57	-64	-71	-78	-84
45	26	19	12	5	-2	-9	-16	-23	-30	-37	-44	-51	-58	-65	-72	-79	-86
50	26	19	12	4	-3	-10	-17	-24	-31	-38	-45	-52	-60	-67	-74	-81	-88
55	25	18	11	4	-3	-11	-18	-25	-32	-39	-46	-54	-61	-68	-75	-82	-89
60	25	17	10	3	-4	-11	-19	-26	-33	-40	-48	-55	-62	-69	-76	-84	-91

Wind Speed (miles per hour) — leftmost column

Frostbite Times (in minutes)   ☐ 30   ☐ 10   ■ 5

Table 2.2
Wind Chill Chart °F *

*Table data is compiled by the National Weather Service.

2. How can you have an actual temperature of 40° and yet an apparent or sensible wind chill temperature near freezing point?

3. How does wind speed influence the apparent or sensible temperature in the following?

   A. In a Michigan Winter? _____

   B. In a Florida Summer? _____

   C. In an Arizona Summer? _____

Just as wind influences the way we perceive temperatures below 50°F, so does humidity influence the way we feel high temperatures.

4. Use the Heat Index Chart (Table 2.3) and calculate the Heat Index for the following:

   A. Actual temperature 85° F, Relative Humidity 60% _____ °F

   B. Actual temperature 95° F, Relative Humidity 70% _____ °F

   C. Actual temperature 110° F, Relative Humidity 50% _____ °F

Air Temperature (°F)

	70	75	80	85	90	95	100	105	110	115	120
30	67	73	78	84	90	96	104	113	123	135	148
35	67	73	79	85	91	98	107	118	130	143	
40	68	74	79	86	93	101	110	123	137	151	
45	68	74	80	87	95	104	115	129	143		
50	69	75	81	88	96	107	120	135	150		
55	69	75	81	89	98	110	126	142			
60	70	76	82	90	100	114	132	149			
65	70	76	83	91	102	119	138				
70	70	77	85	93	106	124	144				
75	70	77	86	95	109	130					
80	71	78	86	97	113	136					
85	71	78	87	99	117						
90	71	79	88	102	122						
95	71	79	89	105							
100	72	80	91	108							

(Relative Humidity on vertical axis)

Heat Index	Affects on the Human Body
130 or above	Heat stroke likely with continued exposure
105 to 130	Heat stroke likely with prolonged exposure
90 to 105	Heat stroke possible with prolonged exposure

Table 2.3 and Table 2.4
Heat Index Chart °F and Heat Index Symptoms*
*Table data is compiled by the National Weather Service.

Section VIII   Isotherms

1. On Figure 2.3 all points of equal temperature are connected by an isotherm. What is the isotherm interval? _____

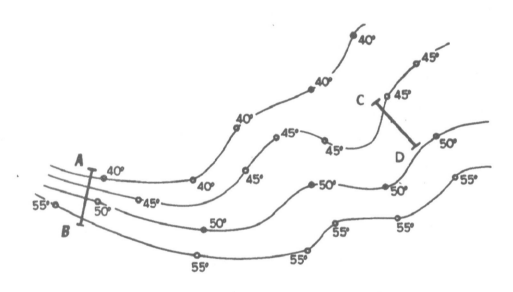

Figure 2.3
Isotherms

2. Is it possible for isotherms to cross each other? _____

3. On the left side of Figure 2.3, the temperature change over the distance A-B is a little more than _____ °F. On the right side of this same diagram, the temperature change from C-D is approximately _____ °F. The change in temperatures over a given distance, such as A-B or C-D is called the _____. Whenever the change is relatively slow over a given distance, it is said to be a (gentle? steep?) gradient. Which distance (A-B or C-D) illustrates a steep temperature gradient? _____

Section IX Carbon Dioxide, Greenhouse Gases and Temperature

Use Figures 2.4 and 2.5 to complete the following questions.

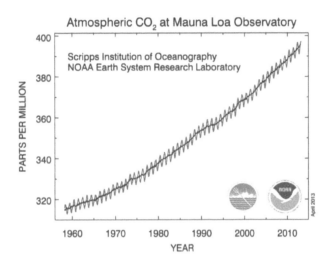

Figure 2.4 1960-2010 Increases in Carbon Dioxide Levels

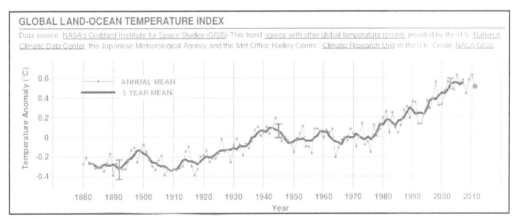

Figure 2.5 1880-2010 Global Land-Ocean Temperature

1. What is happening to the Carbon Dioxide (CO_2) levels in the atmosphere?

2. What sources are causing increases of CO_2 and CH_4 (methane) in the atmosphere?

3. What is happening to the atmospheric temperature from 1880 to 2010?

4. How is CO_2 removed from the atmosphere?

5. What are possible effects of continued CO_2 or CH_4 increases on:

   A. The World's Weather

   B. Ice caps and mean sea level

   C. Agricultural Production

Atmospheric Pressure and Winds   Unit 3

Section I   Pressure Scales

1. Define atmospheric pressure:

2. Name the three common air pressure scales and give the numeric value of the average atmospheric pressure at sea level for each.

    a.

    b.

    c.

3. Name two principle types of pressure measurement instruments and describe their operation.

    _____   _____

                         _____

                         _____

    _____   _____

                         _____

4. Which of the pressure instruments records as a continuous trace?

5. What is an isobar?

6. Warm surface temperatures are most typically associated with _____ pressure. Why?

7. Cold surface temperatures are most typically associated with _____ pressure. Why?

8. Figure 3.1 represents millibar readings for several stations. Complete the diagram by drawing in isobars.

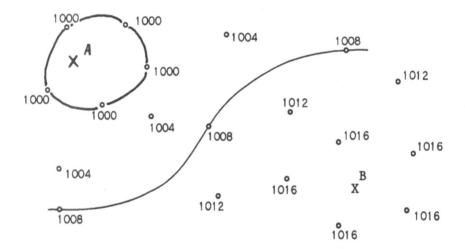

Figure 3.1
Isobars

9. Point A in Figure 3.1 is located near the center of a (high or low) pressure area.

10. Point B in Figure 3.1 is located near the center of a (high or low) pressure area.

Section II   Isobaric Maps

Figure 3.2 depicts a generalized surface pressure map for January 2014. Interpret this map to answer the following questions.

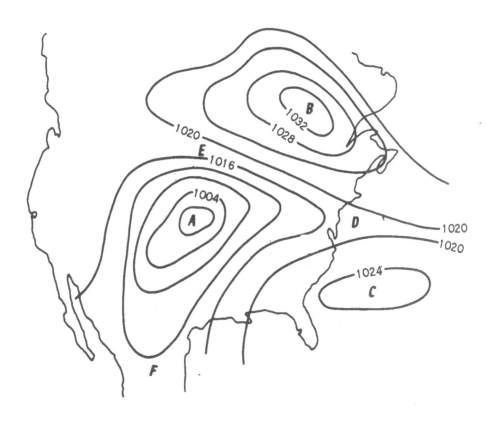

Figure 3.2
Isobaric Map of North America

1. At what point(s) do you observe an anticyclone? _____ and _____

2. At what point(s) do you observe a cyclone?   _____

3. Which point(s) are found in a trough?   _____

4. Which pressure gradient is strongest? B-----A? or C----A? _____

5. What is the value of the isobar closest to point A? _____

6. What is the significance of a steep pressure gradient?

Section III Direction of Air Flow

1. Define wind:

2. Name the three forces which affect the direction of air flow and give a definition of each.

   A.

   B.

   C.

3. On Figure 3.3 indicate the direction of air flow for a cyclone and an anticyclone in the Northern and Southern Hemisphere, taking into consideration Coriolis force and surface friction.

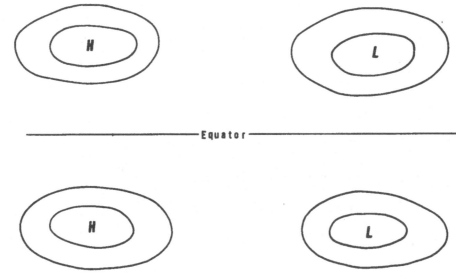

**Figure 3.3
Direction of Air Flow**

Section IV Wind Direction and Wind Speed

Wind direction is frequently confused because of the contradictory uses of directional symbols. Winds are named after the source direction from which they blow. Hence a wind blowing from West to East is known as a West or Westerly wind. This is different from noting it as an Eastward wind (as in toward; this reference is more appropriate for situations such as "driving Eastward [in a car]" or "flying Northward"), and the two are easily confused. A wind vane always points into the source of the wind. Arrows on a map always point along the direction of wind flow -- an arrow may visually indicate Eastward movement, but is still correctly noted as being Westerly.

1. On Figure 3.4 identify the wind direction at the following points.

    A. _____

    B. _____

    C. _____

    X. _____

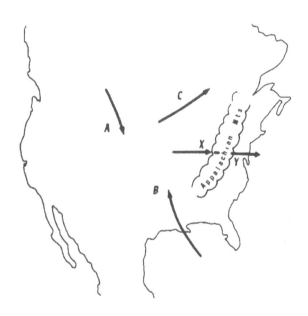

Figure 3.4
Wind Flow Patterns over North America

2. Point X in Figure 3.4 is located on the (windward or leeward) side of the Appalachian Mountains; point Y is located (windward or leeward) slope of the Appalachian Mountains.

3. Name the instrument that measures wind direction _____

4. Name the instrument that measures wind velocity _____

Section V Primary or Global Pressure and Wind Belts

Figure 3.5 represents a view of Earth from a point above the equator. Complete the diagram according to the following instructions.

1. Label the Equatorial Low Pressure and Subtropical High Pressure belts.

2. Draw in arrows to represent the Trade Wind Belt and the Westerly Wind Belt.

3. Label the Polar High Pressure and Subpolar Low Pressure belts.

4. Draw in arrows to represent the Polar Easterlies.

5. After labeling all pressure and wind belts, add the Hadley Cell.

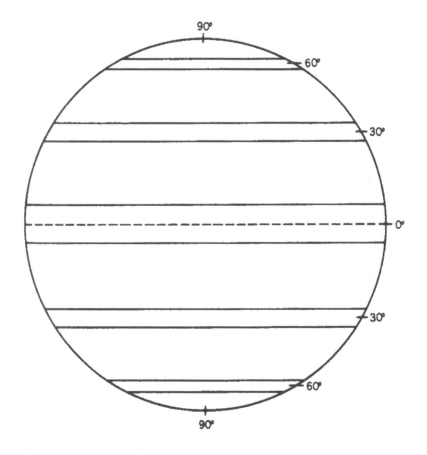

**Figure 3.5
Generalized Primary Pressure and Wind Belts**

6. Since Figure 3.5 represents a generalized model as it might exist during an equinox, discuss what might happen to all the pressure and wind belts during summer solstice.

28

Section VI   Secondary and Regional Pressure and Winds

Secondary or regional pressure and wind systems are generally superimposed on primary pressure and wind belts. They usually occur at the juncture of primary pressure systems and influence large segments of Earth's surface. Two such regional systems are the monsoons and the midlatitude cyclones (discussed in unit 6).

1. Give a brief definition of a **monsoon**.

    A. What causes the seasonal shift of wind direction?

    B. In which areas of the world are monsoons most pronounced?

Section VII   Tertiary or Local Pressure and Wind Systems

As the name suggests, local winds are unique to relatively small geographic regions. Hundreds of such winds are known around the world, often referred to by their local names, e.g. Santa Ana, Chinook, Foehn, Mistral, etc. In this section we will look at important local wind types.

1. Figure 3.6 represents a coastline. Complete it by sketching and/or describing a sea breeze:

    A. At what time of day are sea breezes most common? Why?

    B. During a sea breeze, surface temperatures are (warmer or cooler) over the land as compared to the water.

    C. As a result, atmospheric pressure is (higher or lower) over the land surface, and the wind is blowing from _____ to (show by arrow)..

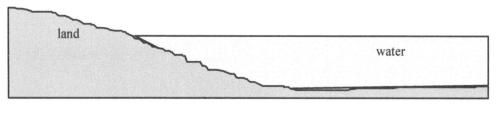

Figure 3.6
Sea Breeze

2. What is a Santa Ana wind? How does it form? Which part of the United States is it most pronounced?

3. What are gravity or Katabatic winds? How do they originate? Name several examples of gravity winds.

4. Sketch or describe Mountain-Valley winds.

Section VIII Experiencing Winds in Different Locations

These questions require thought, and may be made easier because of personal experiences. In either case, looking at a physical globe or map will help.

1. Explain the presence of sand dunes on the eastern shores of Lake Michigan, rather than the western side.

2. Explain why tropical systems (hurricanes) do not directly affect California.

3. What are the primary causes of aridity in the Saharan Desert of Africa, the Sonoran Desert of NW Mexico, or the Atacama Desert of N. Chile? You may need to look at a map or the globe to help you answer this question.

4. Friends of yours are planning a hot-air balloon trip from Chicago, IL to San Francisco, CA. What advice would you give them?

5. What is El Niño? How is the United States affected by El Niño?

Atmospheric Moisture and Precipitation — Unit 4

Section I   Atmospheric Moisture Terms

Define the following terms:

1. **Hydrologic Cycle:**

2. **Evaporation:**

3. **Condensation:**

4. **Precipitation:**

5. **Dew Point Temperature:**

6. When air at given temperature holds all of the water vapor that is possibly can hold, air is said to be in a state of _____ and has reached its _____.

7. In order for condensation to happen, temperatures must **(cool, warm)**.

8. Air's capacity to hold more water vapor **(increases, decreases)** with increasing air temperature.

9. The air temperature is 50°F and the Dew Point Temperature is 38°F. What does the air temperature have to be in order for the air to be saturated? _____

10. What does it mean when the Relative Humidity is 100%?

Section II   Measuring Relative Humidity

Relative humidity varies with the temperature of the air and the actual water vapor content. The formula used to measure relative humidity is

$$R.H. = \frac{S.H.}{M.S.H} \times 100$$

S.H. (specific humidity) refers to the amount (grams) of water vapor present in a unit weight of air, and **M.S.H. (maximum specific humidity)** refers to the maximum amount of water vapor which a unit of air can hold at a specified temperature. Multiplying this ratio by 100 gives us the R.H. (relative humidity) in percent. Measurements of relative humidity are made using the Relative Humidity Chart.

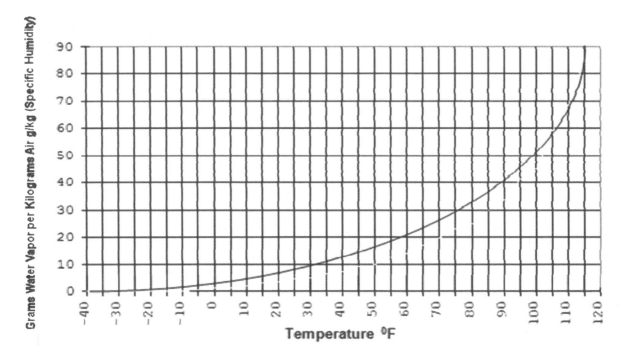

Figure 4.1
Maximum Specific Humidity

Use Figure 4.1 to answer the following questions. Write out the formula when doing calculations.

1. The air temperature is 100°F. The specific humidity is 15 grams per kg.

   A. What is the relative humidity?

   B. What is the approximate dew point temperature?

2. The air temperature is 50°F and the relative humidity is 100 %: What is the specific humidity?

3. Find the Relative Humidity for the following: Write out the formula when doing the calculation.

   A. Air Temperature = 80°F and Specific Humidity = 10 g/kg

   B. Air Temperature = 60°F and Specific Humidity = 10 g/kg

   C. Air Temperature = 40°F and Specific Humidity = 10 g/kg

   D. Air Temperature = 30°F and Specific Humidity = 10 g/kg

   E. The air temperature is 70°F and the specific humidity is 20 g/kg. What is the Dew Point Temperature?

   F. Explain the relationship between Relative Humidity and Temperature when Specific Humidity stays the same. Notice this relationship for questions A-D.

4. What is your comfort level if the dew point temperature is:

   63°? _____

   78°? _____

   57°? _____

   72°? _____

Section III Humidity, Condensation and Precipitation

A diurnal temperature-relative humidity chart of Tucson, Arizona is shown ion Figure 4.2.

1. In your own words describe the temperature and relative humidity relationship for a single day.

2. State whether conditions in Tucson are favorable for the formation of dew.

3. What are the probabilities of dew occurring the next night?

4. Explain how ground temperature is important in dew formation.

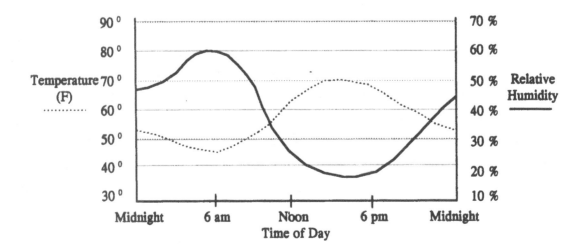

Figure 4.2
Adapted from Battan, Weather in Your Life, 1983, p12

5. If the base of a cloud is at a very low altitude of 1,000 ft, is suggests that only a few degrees of cooling was necessary to reach the condensation level and that the air near the ground was close to _____ or near _____ R.H.

6. If the relative humidity is 100 % near the ground, what type of condensation might you expect?

   _____

7. It is raining where you are, yet the relative humidity is only about 75 % near the ground. Give an explanation for this.

Section IV Atmospheric Lifting an Cooling

1. Make three sketches (showing the location of clouds and precipitation) that represent:

    A. Orographic lift

    B. Convectional lift

    C. Frontal lift

2. Which source of lift is primarily responsible for precipitation in the tropics?

3. Which source of lift is primarily responsible for precipitation in the mid-latitudes (such as 40°N)?

4. Which source of lift is primarily responsible for precipitation in mountain regions?

Section V Stability and Instability

1. On Figure 4.3 plot the average environmental lapse rate (3.5°F) (dashed line - - - - - - - -) and the dry adiabatic lapse rate (solid line _____), from a surface temperature of 70°F. Stop at 6,000 feet. Compare the air temperatures at several given heights.

   Given the situation above, we can state that air rising and cooling at the dry adiabatic rate is at all times (**warmer or cooler**) than the air through which it is passing. Hence it will (**rise or sink**) and the atmosphere is said to be (**stable or unstable**).

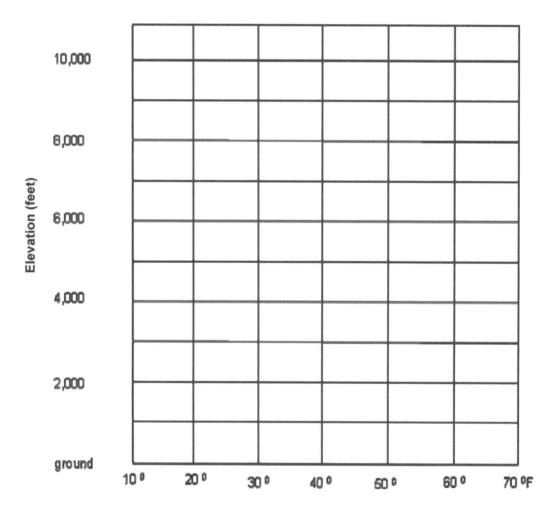

Figure 4.3
Stability and Instability

2. In a second example, use the same dry adiabatic rate as above and add the environmental lapse rate of rate of 8°F / 1,000 ft (with a dotted line ....). Stop at 6,000 feet. Compare the temperatures at given heights.

   Given this situation of a lapse rate of 8°F / 1,000 ft, we can state that air rising and cooling at the dry adiabatic rate will at all times be (**warmer or cooler**) than the surrounding air. Hence it will tend to (**rise or sink**) and the atmosphere is said to be (**stable or unstable**).

Section VI Cloud Types and Cloud Identification

1. In Table 4.1, briefly describe the most probable weather conditions when you observe the following clouds during the summer.

Cloud Type	Altitude (ft)	Associated Weather Characteristics (e.g. precip., sunshine)
Cirrus		
Cirrostratus		
Cirrocumulus		
Altostratus		
Altocumulus		
Stratus		
Stratocumulus		
Nimbostratus		
Cumulus		
Cumulonimbus		

Table 4.1
Characteristics of Summer Cloud Types

2. Identify the clouds in the pictures below. Include the cloud's height.

A. Name: Height: B. Name: Height:

C. Name: Height: D. Name: Height:

E. Name: Height: F. Name: Height:

G. Name: Height: H. Name: Height:

Section VII Types of Fog

1. Complete Figures 4.4, 4.5, and 4.6. Give a brief explanation to show how radiation, advection, and evaporation fogs are formed.

**Figure 4.4
Radiation Fog**

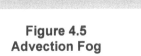

**Figure 4.5
Advection Fog**

**Figure 4.6
Evaporation Fog**

Section VIII Forms of Precipitation

1. Complete Figures 4.7, 4.8, 4.9, and 4.10. Name the type of precipitation represented in each diagram. Use the weather symbols in Figure 4.11 to illustrate the form of precipitation as it falls from the cloud to the surface.

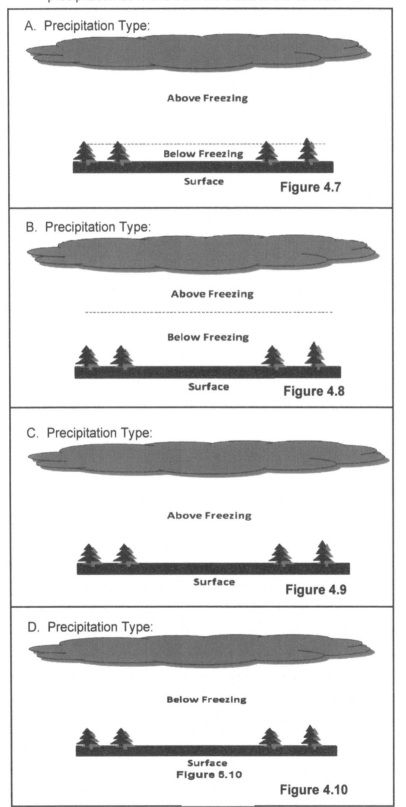

Figure 4.11

Section IX Cloud Picture and Identification   Name:

I would like you to do the following for one cloud:

1. Observe the sky! Take a picture of the cloud.
   Please give the following information below for the cloud:
   - Name of the cloud and time, date, temperature, dew point temperature, wind direction and location of where the picture was taken.
   - Define the cloud (include height). Use your notes, book, or online source.

2. Being There: Sense of Place Inventory – next page

Name of Cloud: _____
Location: _____
Date and Time: _____
Temperature: _____
Dew Point Temperature: _____
Wind Direction:_____
Definition and Picture:

Being There: Sense of Place

Direct experience of a landscape immerses us in its physical and human characteristics, spatial arrangements, environmental qualities, sights, sounds, textures, taste, and smells. The immersion experience of landscape evokes a wide range of feelings and thoughts. Thoughts revolve around our personal conceptual framework. Feelings emerge from our emotional state. Our experience is a mixture of objective and subjective realities.

Location: _____

Below, please provide a description of the landscape you are currently experiencing with any or all senses. What is there? Where is it? How is it arranged? How did it make you feel? Why is it important to you?

_____

_____

_____

_____

_____

_____

_____

_____

_____

_____

_____

_____

_____

_____

_____

_____

_____

_____

_____

_____

Mid-Latitude and Tropical Weather Unit 5

Section I   Air Masses

1. On figure 5.1, label all air mass types affecting the continental United States using abbreviated symbols.

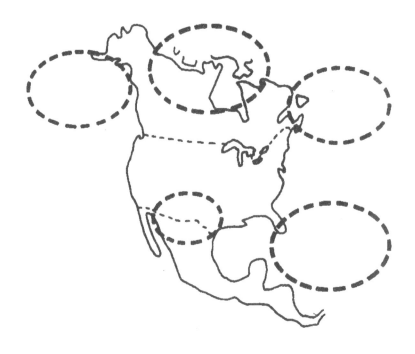

Figure 5.1
Air Mass Source Regions

2. Define **air mass**.

3. Polar continental air masses are associated with what type of weather conditions?

4. A tropical maritime air mass that moves from the ocean onto a warm land surface is likely to produce (stable or unstable) weather conditions. Explain.

5. A polar maritime air mass that moves from the ocean onto a warm land surface is likely to produce (stable or unstable) weather conditions. Explain.

6. Which two air masses most commonly influence the weather of Illinois? Why?

Section II Mid-Latitude Cyclones

1. Complete the diagram in Figure 5.2 to represent a weather-map view of a mid-latitude cyclone, at a mature stage, in the northern hemisphere. Include:

   A. area of lowest pressure

   B. cold and warm front

   C. mT and cP air masses

   D. wind directions

**Figure 5.2
Northern Hemisphere Cyclone**

2. Figures 5.3 and 5.4 represent cross sections of a warm front and a cold front. Complete the diagrams by (a) identifying the type of front, (b) labeling the clouds associated with each front, and (c) the cold and warm air and the direction the air is heading.

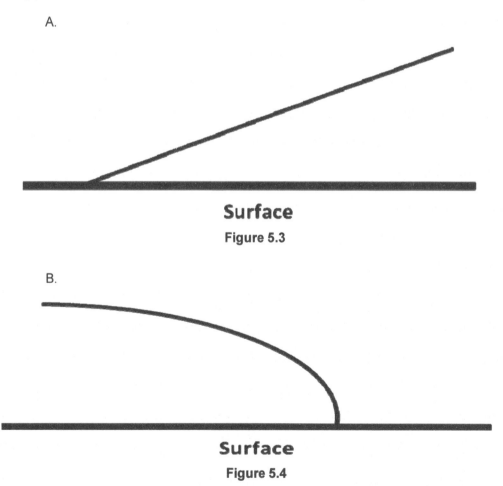

46

Section III Weather Maps

Figure 5.5 shows a station model which gives the legend to a modern digital weather map, or surface analysis map. Note that information given around any reporting station is in a specific shorthand form. Familiarize yourself with symbols of pressure, temperatures, winds, precipitation, and cloud cover. Note that if the first digit of the pressure is 0-5, you add 10 in front and a decimal before the last digit (147 = 1014.7 mb). If the first digit is 6-9, you would add 9 in front and a decimal before the last (892 = 989.2 mb). Figure 6.4 shows a weather map using that legend. Refer to both figures to answer the following questions.

**Figure 5.5
Station Model**

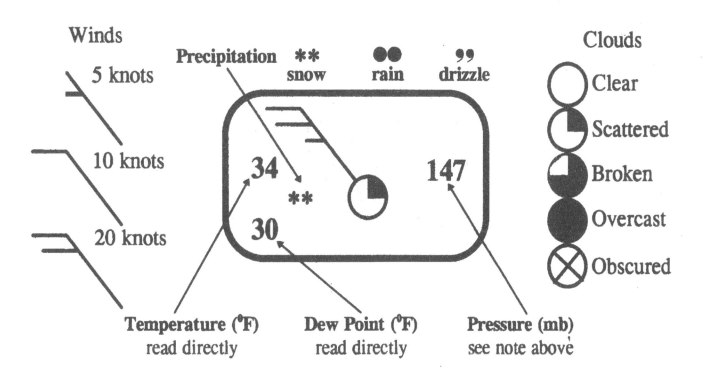

Using the weather map (Figure 5.6) at the end of the unit, answer the following questions.

1. Record the missing information, in the following table, for the listed reporting stations.

	Air Temp (°F)	Dew Point Temp (°F)	Wind Dir.	Wind Spd. (mph)	Pressure (mb)	Clouds (%)
Des Moines, IA						
Little Rock, AR						
Atlanta, GA						
Austin, TX						

2. Is the relative humidity higher in Atlanta, GA or Austin, TX? Why?

3. What is the wind speed in Reno, Nevada (north near CA border)? _____

4. Which state(s) is/are experiencing the heaviest precipitation? Why?

5. The center of the major cyclone on this map is located over which state? _____

6. What is the pressure near the center of this cyclone? _____

7. Is this cyclone showing signs of occluding? _____

8. Note the position and alignment of the fronts associated with the cyclone. Through what states does the warm front pass?

9. Through which states does the cold front pass?

10. What air mass type lies to the **south and east** of the center of the cyclone? _____

11. What air mass type lies to the **north and west** of the center of the cyclone? _____

12. What type of a front is located in the northern part of Montana? _____

13. What type of front is located in the Pacific Ocean near the Low Pressure? _____

14. Knowing that this cyclone is moving in a NE direction at a speed of 30 miles per hour, give the current weather conditions for Detroit, MI for April 3, 2014 and a general weather forecast for Detroit, Michigan, for April 4, 2014.

	April 3	April 4
A. Temperature		
B. Pressure		
C. Wind Direction		
D. Chance of Precipitation		

Section III Weather Maps Part II

Use the blank surface weather map (Figure 5.7) to draw weather data for the reported stations. Once the station models are drawn, add the warm front and the cold front according to the weather station data.

1. Draw the station models on the weather surface map (Figure 5.7).

A. Milwaukee, WI:
   Temperature: 50°F
   Pressure: 996.0 mb
   Dew Point: 50°F
   Wind Direction: SE, 15mph
   Degree of Cloudiness: 100%, showers

B. Cincinnati, OH:
   Temperature: 58°F
   Pressure: 1000.0 mb
   Dew Point: 58°F
   Wind Direction: SE, 10mph
   Cloudiness: 100%, showers

C. Indianapolis, IN:
   Temperature: 76°F
   Pressure: 1008.0 mb
   Dew Point: 70°F
   Wind Direction: SW, 20mph
   Degree of Cloudiness: 25%

D. Springfield, IL:
   Temperature: 75°F
   Pressure: 1010.0 mb
   Dew Point: 69°F
   Wind Direction: SW, 20mph
   Cloudiness: 25%

E. Kansas City, MO:
   Temperature: 70°F
   Pressure: 1004.0 mb
   Dew Point: 69°F
   Wind Direction: South, 30mph
   Degree of Cloudiness: 100%, Thunderstorms

F. Omaha, NE:
   Temperature: 38°F
   Pressure: 1020.0 mb
   Dew Point: 20°F
   Wind Direction: NW, 5mph
   Cloudiness: 0%, clear

2. The center of lowest pressure is 998 millibars and is located in Minneapolis, MN. Label the center of low with an **L** on the map.

3. Draw the approximate location of the warm front.

4. Draw the approximate location of the cold front.

Section IV   Severe Thunderstorms

1. A thunderstorm is classified as severe when what conditions occur?

2. What is a supercell thunderstorm?

3. What is a microburst?

4. What is heat lightning?

5. Give three lightning safety tips.

Section V   Tornadoes

1. What is the difference between a Tornado Watch and a Tornado Warning?

2. How is the magnitude of the Fujita scale determined?

3. The majority of tornadoes are categorized under which Fujita scale(s)?

4. Which state has the highest annual number of tornadoes?

5. Explain physical geographic factors which account for the high occurrence of tornadoes in the American Midwest and Great Plains. (Hint: Identify topography and air masses.)

6. With which weather front are tornadoes most associated?

7. Give three tornado safety tips.

Section VI   Tropical Disturbances

Give several reasons why tropical hurricanes originate where they do and why they dissipate over land areas.

1. How do hurricanes differ from mid-latitude cyclones for the following factors?

   A. Wind Velocities

   B. Wind Direction

   C. Temperature Distribution

   D. Direction of Travel

   E. Pressure Gradient

   F. Geographic Location

Figure 5.6 American Meteorological Society Surface Weather Map

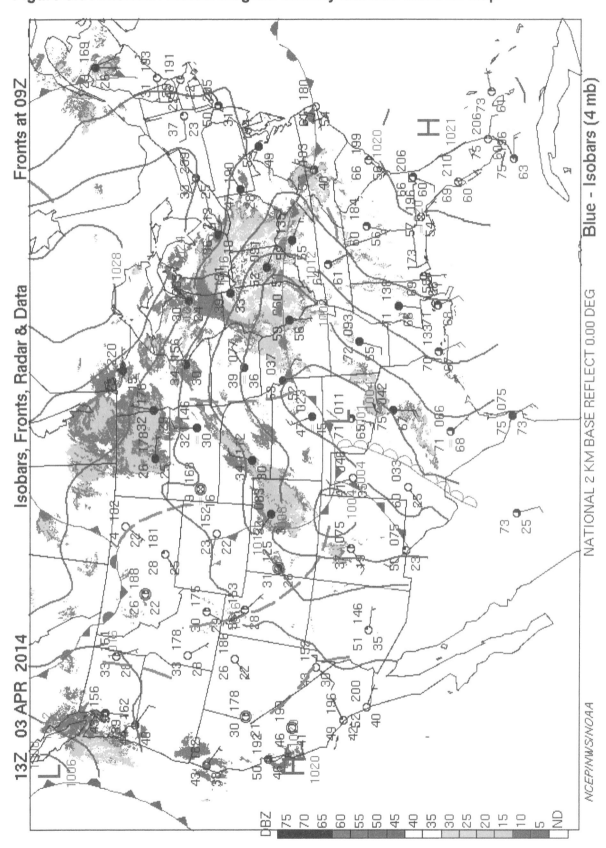

Figure 5.7 Surface Weather Map

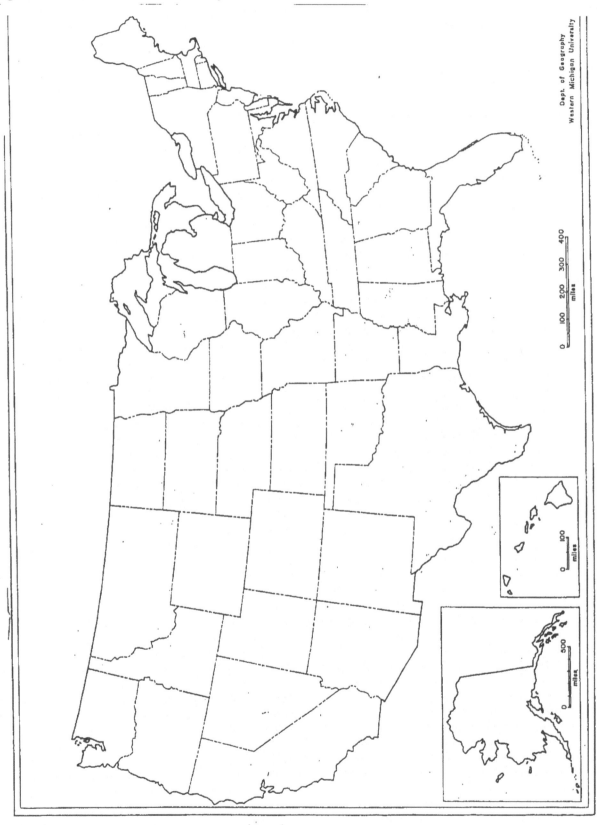

Maps and Map Interpretation   Unit 6

Section I   USGS Topographic Quadrangle Maps

The user of topographic maps should have a working knowledge of standardized map symbols on U.S. Geological Survey (USGS) topographic quadrangle maps.

1. What are the five basic colors used on the USGS series of topographic quadrangle maps and what general categories are represented by each color?

	Color	Meaning
A.	_____	_____
B.	_____	_____
C.	_____	_____
D.	_____	_____
E.	_____	_____

2. On the more recent topographic maps the color **purple** is used. What is its function?

3. Draw the symbols for the features indicated in the list below. (If colors other than black and white are needed, either color the symbol or simply make a note stating the color of that symbol.)

Symbol	Feature	Symbol	Feature
_____	Mine or quarry	_____	Road bridge
_____	Intermittent streams	_____	School
_____	Perennial streams	_____	Church
_____	Woods or brush woods	_____	Cemetery
_____	Dry Lake	_____	Marsh or swamp
_____	Railroad, single track		

4. Examine the Divernon Quadrangle. Name a feature on the quadrangle that is purple.

Section II   Measuring Distance

1. A. Using the USGS 7.5 minute Topographic Quadrangle map excerpt for Divernon, IL,

   Representative Fraction   _____

   Written or Verbal Scale   _____

   B. Measure the Illinois Central Railroad distance from the northern edge to the southern edge of this map.

   _____ miles

   C. Measure the distance of Panther Creek in the northwest corner of the map.

   _____ miles

   D. A dashed black line indicates the city limits of Divernon. How large (in area) is the city of Divernon? (Area = Length X Width)

   _____ sq miles

2. How far is it to:

   A. Springfield, if you take Route 4?   _____

   B. Springfield, if you take Interstate 55?   _____

   C. Litchfield, if you Interstate 55?   _____

Section III   Map Features

1. Name the counties found on the Divernon Quadrangle?

2. Name the cemeteries found on the Divernon Quadrangle.

Section IV   Map Scales

1. Convert the following representative fractions to verbal scales.

    A. RF of 1:63,360 to written scale **miles-to-the-inch** _____

    B. RF of 1:250,000 to written scale **miles-to-the-inch** _____

    C. RF of 1:1,000 to written scale **miles-to-the-inch** _____

2. Express the following verbal scales as representative fractions.

    A. Written scale **2 inches-to-one-mile**   1:_____

    B. Written scale **1 inch-to-250 miles**   1:_____

    C. Written scale **3 inches-to-6 miles**   1:_____

3. If a map 12 by 12 inches and at an RF of 1:62,500 is redrawn at an RF of 1:125,000, this map would have the following dimensions: _____ x _____ inches

4. Use the three maps selected by your instructor and give their **representative fractions**.

    A. Map #1 _____

    B. Map #2 _____

    C. Map #3 _____

    D. Which of the three maps has the largest scale? _____

5. Explain the difference between a "large scale" map and a "small scale" map.

Section V   Township and Range

1. Figure 6.1 represents a township. Assume that each square below is a section of land as defined by the United States Land Survey. On the first line for Section 3 you will find the correct description of land parcel X in that section. Complete the description for parcel X, Y and Z for each of the three indicated sections below.

Section 3, parcel:
X _W ½, SE ¼_____
    Area 80 acres
Y _____
    Area _____ acres
Z _____
    Area _____ acres

Section 17, parcel:
X _____
    Area _____ acres
Y _____
    Area _____ acres
Z _____
    Area _____ acres

Section 26, parcel:
X _____
    Area _____ acres
Y _____
    Area _____ acres
Z _____
    Area _____ acres

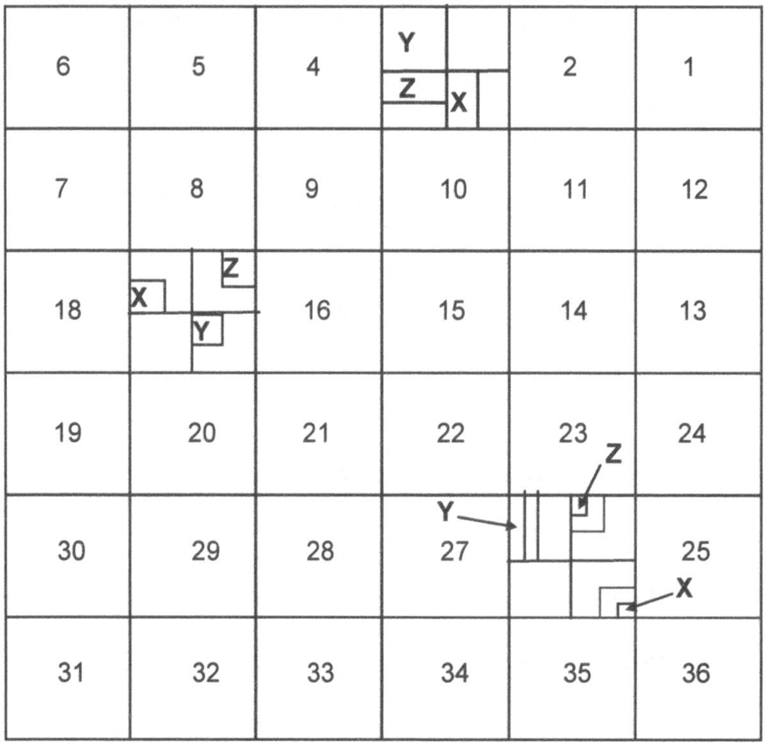

T 26 N

R 75 W

Figure 6.1
Township

2. On Figure 6.2, outline, label (by letter), and shade in the parcels of land described in questions a, b, and c.

Figure 6.2
Section 21

a. NW ¼ , SW ¼ , Sec. 21 _____ acres
b. E ½ , E ½ , NE ¼ , Sec. 21 _____ acres
c. NW ¼ , NW ¼ , NW ¼, Sec. 21 _____ acres
d. S ½ , N ½ , SE ¼ , Sec. 21 _____ acres

3. List the names of the townships found on the Divernon, IL topographic map.

4. Use the Divernon, IL topographic map and identify the cultural feature in the SW ¼ , NW ¼, Sec. 32, T. 14 N., R. 5 W.
   _____

5. Give a detailed legal (Land Survey System) of the sewage disposal plant approximately one mile east of the City of Auburn.
   _____

6. Give a detailed legal (Land Survey System) of the quarry approximately one mile north of the City of Divernon.
   _____

Section VI   Metes and Bounds Survey System

Prior to the establishment of the Land Survey System, property ownership relied on a survey system known as Metes and Bounds. **Bounds** are the external or limiting lines of any object or space. These could be trees, streams, cliffs, wells or other physical features that assisted in defining the shape of the property. **Metes** are the descriptions of adjacent property features expressly excluded from that described, and frequently indicated the ownership of adjacent property. This system was used in the original thirteen colonies and seven other states. Notice the lack of consistent pattern of property lines in a metes and bounds survey as compared to the rectangular pattern aligned north-south and east-west in the Land Survey System.

Figure 6.3

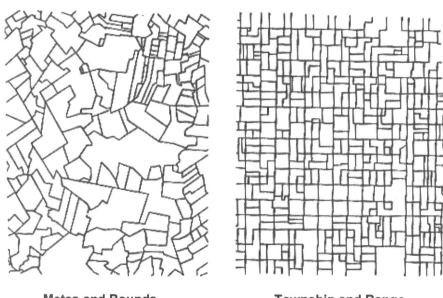

Metes and Bounds   **Township and Range**

Section VII Contour Lines

A contour line is defined as a type of isoline, or imaginary line, which connects points of equal elevation on the surface of Earth. In other words, each point on the same contour has the same elevation. Contour lines appear on the topographic map series of every country which publishes such maps, and the basic principles of contour mapping can be applied to any data, even if the units of measurement are different. After reading the descriptive text provided in this lab book, and examining the Afton, Iowa quadrangle, and available raised-relief maps, indicate whether the following statements are **True** or **False**.

1. _____ All topographic quadrangle maps are published at the same scale.

2. _____ Contour lines are used to show cultural features.

3. _____ Contour lines are printed in brown.

4. _____ Contour lines can intersect each other.

5. _____ Contour lines tend to bend, or point downstream.

6. _____ The contour interval is the vertical distance between contour lines.

7. _____ Contours are spaced equally apart when the slope of the ground is uniform.

8. _____ Numbered contour lines are the only indications of elevation on a topographic map.

9. _____ Contour lines which are closely spaced indicate an area of relatively steep slopes.

10. _____ The same contour interval is used on all topographic maps.

Section VIII Elevation and Local Relief

In working with topographic maps, a distinction should be made between the terms elevation and relief. The term **elevation** refers simply to the height of a place above some datum plane, usually sea level. On the other hand, **local relief** is the vertical difference, measured in feet, between two elevations within a given area. Thus, a certain mountain may have an elevation of 3,000 feet above sea level; however, if the lowest part of the area is 2,000 feet and the highest is 3,000 feet, the local relief of this area is 1,000 feet.

Determining the elevation of a given point on a map may be done in several ways. First, there are many places for which elevations are given; on the Afton Quadrangle, you will observe a number of **bench marks** (sometimes indicated by the symbol BM), as well as elevation numbers at almost every road intersection. The elevation of any other point can be determined by **interpreting or interpolating the contour lines**. If a point is exactly located on a contour line, its elevation can be readily determined; for example on Figure 6.4 point D is at an elevation of 340 feet. If a point is located between contour lines, its elevation can be approximately determined. For example, in Figure 6.4, point K is located between the 320-foot contour and the 340-foot contour; i.e., the elevation is above 320 feet but below 340 feet. Therefore, point K is said to be at an elevation of between 320'-340'. The exact elevation of point K can only be suggested but never be accurately determined on this map. Similarly, point H has an elevation of between 420'-440'.

Using the Divernon, IL Quadrangle, work the following problems:

1. What is the contour interval of the Divernon Quadrangle?

2. Determine the elevation of the Auburn Cemetery found in Sec. 15, T.13 N., R.6 W.

3. Determine the elevation of the Sugar Creek Church found in Sec. 31, T.14 N., R.5 W.

4. Determine the elevation of the City of Divernon.

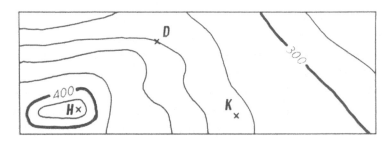

Figure 6.4
Contour Map

The Earth's Crust and Earth Materials Unit 7

Section I   Earth's Major Layers

1. Earth's solid sphere can be divided into three major divisions, separated according to chemical and physical differences. Describe briefly each sphere as to thickness, its probable composition, and its rigidity.

   A. Crust

   B. Mantle

   C. Core

2. The crust is usually separated into two segments. Name them and describe the composition of each.

   A. _____

   B. _____

3. Describe the location of the asthenosphere and explain its significance.

4. Draw a model of Earth (include the inner core, outer core, mantle, crust, and distances).

Section II   Elements of the Earth's Crust

The composition of the Earth's crust can be studied in terms of its **elements**, **minerals**, **rocks**, and **landforms**.

1. **Elements**, or types of atoms, are the most basic building blocks of all matter on Earth. Of the approximately 100 naturally occurring elements, list the five that are most abundant on Earth's surface.

    _____

    _____

    _____

    _____

    _____

2. **Minerals** are usually composed of one or more elements. Each mineral has a unique chemical composition and physical structure and thus, can be readily identified. The great bulk of rock-forming minerals on the Earth's crust are either made up of quartz (silicon dioxide) or other silicate minerals (minerals in which the quartz has combined with other elements such as calcium, sodium, magnesium, etc.).

    A. List several varieties of quartz.

    B. List four common silicates and place them in order of increasing density.

    _____

    _____

    _____

    _____

    C. **Silicates** are only one of several distinct families of minerals; others are **oxides** and **carbonates**. Name two examples of each of these.

    Oxides _____   _____

    Carbonates _____   _____

3. **Rocks** are aggregates of minerals in various proportions. Depending on where and how the rocks originate in their geological history, they are placed into three major classes. Name them and describe how each class is formed.

   _____

   _____

   _____

Section III Igneous Rocks

1. All igneous rocks are formed from magma. If the magma cools and solidifies very slowly, as it does deep within Earth's crust, then intrusive igneous rock is formed.

   A. Describe the characteristics of intrusive igneous rocks.

   B. Give several examples of intrusive igneous rocks.

2. Magma which flows onto Earth's surface cools quickly and solidifies to form extrusive igneous rock.

   A. How does extrusive rock differ from intrusive igneous rock?

   B. Give several examples of extrusive igneous rocks.

   C. Give the intrusive equivalents of the extrusive rocks you listed above.

Section IV   Sedimentary Rocks

1. How are clastic sedimentary rocks formed?

2. Name three clastic sedimentary rocks in decreasing order of particle size.

   _____

   _____

   _____

3. Give two examples of chemically precipitated sedimentary rock.

   _____

   _____

4. Give two examples of organic sedimentary rocks.

   _____

   _____

5. What is an evaporite?

Section V   Metamorphic Rocks

1. How are metamorphic rocks formed?

2. Give the metamorphic counterpart of each of the following rocks.

   Granite _____

   Sandstone _____

   Limestone _____

   Shale _____

Section VI   Rock Recognition

In studying landscapes in the field, it is necessary to recognize some of the common rock types and to know something about their properties. Rocks can be identified by a number of properties including color, texture, mineral composition, chemical reaction to acid, and mode of occurrence. The information in Table 7.1 will serve as a guide for identifying some common rock types.

Rocks are frequently grouped into three main classes based on their mode or origin: **igneous**, **sedimentary**, or **metamorphic**. Consult your text or other resource for major rock characteristics.

1. Examine the rock specimens provided in the laboratory and identify them based on the information given in the Table 7.1

	Rock Type	**Identifying Characteristics**
1.		
2.		
3.		
4.		
5.		
6.		
7.		
8.		
9.		
10.		
11.		
12.		

Table 7.1 Characteristics of Some Common Rock Types

	Rock Type	Color	Texture	Other Distinctive Properties
Igneous	Granite	Light grey or pink speckled with grey or black crystals	Coarse crystals clearly visible, including quartz, feldspar, and mica	Occurs in large massive bodies
	Rhyolite	Reddish	Fine, crystals not visible	Massive bodies
	Gabbro	Dark grey	Coarse crystals, not clearly visible because all are dark color	
	Basalt	Black or dark grey	Fine, crystals no visible	Columnar structure is massive
Sedimentary	Conglomerate	Varied, but usually tan to brown	Large gravel to sand-size particles imbedded in matrix of sand grains	Rounded fragments
	Sandstone	Varied; usually buff or tan	Sand-sized rounded particles	Frequently in clearly definable beds, 6" to 5' or more thick
	Shale	Varied; usually light to dark grey	Silt to clay size particles	Tends to break up into small thin layers; soft, can be scratched by fingernail
	Limestone	Usually light to grey	Dense fine crystals, or oolitic presence of fossils common, lacks stratification	Fizzes when drop of acid applied
Metamorphic	Slate	Dark, but variable	Fine grained (microscopic)	Highly foliated, will split into thin sheets
	Gneiss	Bands of light and dark minerals	Crystalline similar to granite, foliated	No highly developed cleavage
	Quartzite	Light pink, white, red, grey – shiny	Glassy crystals, non foliated	No highly developed cleavage
	Marble	Variable, but usually white to grey	Non-foliated; impurities produce interesting colors and designs	Does not fizz when acid is applied

The Earth Structure and Tectonic Processes   Unit 8

Section I   Continental Drift and Tectonics

1. For each of the general topics listed below, describe how scientists have found evidence suggesting past and present movement of continental and oceanic crust.

   A. Earthquakes and Fault Zones

   B. Earth's Magnetism

   C. Examination of a Globe or World Map

   D. Mountain Ranges

   E. Fossil Organisms

   F. Climate

2. Based on his own observations, Alfred Wegener proposed a hypothesis of continental drift. The theories have been modified somewhat and are now referred to as Plate _____. How do modern earth scientists envision the movement of continents?

3. What is the difference between an oceanic lithospheric plate and a continental lithospheric plate?

4. What is the probable mechanism or energy source that causes plate motion?

5. What might happen when an oceanic lithospheric plate and a continental plate collide or converge? Draw a simple sketch and label those parts known as the subducting plate, the ocean trench, and the rising magma producing volcanoes.

6. What happens when two plates move apart or diverge? Draw a sketch and label the plates producing sea floor spreading, and the zone of rising magma.

7. Give at least one example of a geographic region where the following movements have taken place.

   A. Collision of continental and oceanic plate (convergent plates)

   B. Collision of two continental plates (convergent plates)

   C. Plate divergence (sea floor spreading)

   D. Transverse movement (horizontal movement of plates)

   E. Collision of two oceanic plates

8. Reconstructing the movement of plates over geologic time, scientists hypothesize that some _____ million years ago, during the _____ period, all the present continents formed one supercontinent known as _____. When this supercontinent broke up in to main segments, the northern hemisphere segment was known as _____ and the southern hemisphere segment was known as _____.

Section II Plutons

Plutons are intrusive igneous bodies of various sizes which originate deep below the Earth's surface. They are often also exposed at the surface after thousands of years of erosion.

1. Label the following features in Figure 8.1.

   A. Batholith

   B. Sill

   C. Dike

   D. Laccolith

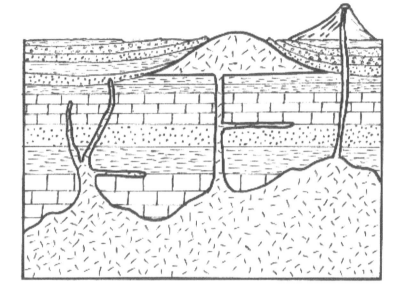

Figure 8.1

2. Which type of pluton is represented by the following geographic places?

   A. The Sierra Nevada

   B. The Black Hills

Section III Plate Fractures, Diastrophism

Collision of two plates leads to much deformation and fracturing of the lithosphere, a process known as diastrophism.

1. Sketch and label all parts of a fold in sedimentary rock.

2. What is a fault and how can you recognize one?

3. Faulting can take place in a variety of ways. Label items A - C in Figure 8.2.

   A. An upthrown block

   B. A downthrown block

   C. A fault scarp

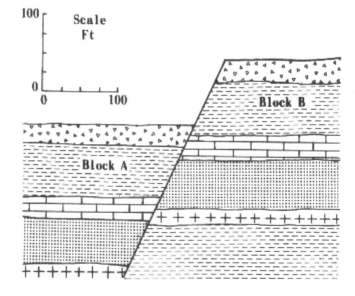

Figure 8.2

4. Figure 8.2 represents a normal or reverse fault?

5. What is the amount (in feet) of vertical and horizontal movement on the fault shown in Figure 8.2?

   A. Horizontal _____ feet

   B. Vertical _____ feet

6. Make a sketch of a horst and graben and label all the parts.

Section IV   Volcanoes

The continuous movement of plates has produced much unrest along the plate boundaries, which is evidenced by the unusually widespread occurrence of volcanism and earthquakes along these zones.

1. Define lava.

2. How does lava differ if it:

    A. has a higher quartz content and originates over continental crustal surfaces?

    B. has a lower quartz content and originates within the oceanic crust?

3. Suggest the type of volcano and lava that might be found at each of the following:

    A. Hawaiian Islands

    B. Iceland

    C. Mount Vesuvius

    D. Mount St. Helens

4. What is tephra and what type of volcano does it form?

5. Describe a composite volcano and name several volcanoes of that type.

6. Describe a shield volcano.

7. Draw a generalized cross section of a composite volcano, shield volcano, and cinder cone in the space below. Use shading to differentiate between any layers of lava, pyroclastic materials, or tephra, and label the common components of each of the types. Indicate in some way the relative sizes of each of these types of cones.

8. How does a caldera form? Give an example of a caldera.

Section V   Earthquakes

1. How are earthquakes generated?

2. What other forces can you name that cause tremors of Earth's surface measurable with a seismograph?

3. The magnitude of ground motion is measured by the _____ scale, and an instrument measuring this is called a _____.

4. An earthquake with a magnitude of 6 is _____ greater in terms of energy released than a quake with a magnitude of 4. An earthquake with a magnitude of 8 is _____ greater in terms of energy released than a quake with a magnitude of 4.

5. Name three regions within the United States where strong and frequent earthquakes are registered.

   _____

   _____

   _____

6. What is a tsunami, and how is it produced?

Weathering and Mass Wasting — Unit 9

Section I   Gradation

1. Define each of the following:

    A. Gradation

    B. Degradation

    C. Aggradation

    D. Weathering

    E. Erosion

    F. Mass Wasting

2. What is the ultimate result of gradation?

3. What are the sources of energy which produce gradation?

4. What forces continuously oppose the ultimate results of gradation?

Section II   Weathering

1. List three physical and three chemical weathering processes and state how each process operates.

   A. Physical Weathering

   _____

   _____

   _____

   B. Chemical Weathering

   _____

   _____

   _____

2. Indicate by letter (P or C) which processes listed below are physical and which are chemical.

Process	Nature of Process	Process	Nature of Process
Unloading		Frost Action	
Crystal Growth		Solution	
Carbonation		Oxidation	
Hydrolysis		Plant Roots, Burrowing, Animals	

Section III   Mass Wasting and Erosion

Mass wasting refers to the downslope movement of materials under the control of gravity. No water or any other medium is needed for this gravity movement, though moisture may be associated with some forms. Some downslope movements proceed at an exceedingly slow rate such as soil creep, others such as landslides or rockfall are devastatingly fast.

1. Under which environmental conditions do the following mass wasting process occur?

   A. Soil Creep

   B. Earth, Mud and Rockflows

   C. Landslide

   D. Rockfall

2. Under what special conditions does solifluction operate?

3. Erosion can be defined as the **"picking up and removal of earth materials other than by gravity"**. List the four most important erosional agents.

   A. _____

   B. _____

   C. _____

   D. _____

4. How do human activities influence the rate of mass wasting and erosion?

Surface and Groundwater Hydrology — Unit 10

Section I   The Hydrologic Cycle

1. Label the sequence of events, in Figure 10.1, in the hydrologic cycle.

    A. Evaporation

    B. Transpiration

    C. Precipitation

    D. Runoff

    E. Infiltration

    F. Groundwater Storage

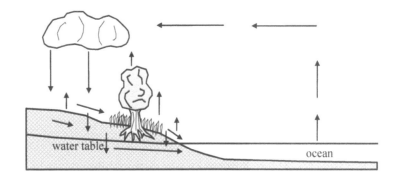

Figure 10.1

2. Differentiate between runoff and base flow in relation to a stream.

3. What happens when the water table is far below a stream bed, thus preventing base flow?

4. What factors determine runoff vs. infiltration?

5. Differentiate between capillary and gravity flow of water in soil.

Section II   Groundwater

1. Define **aquifer**.

2. Define **aquiclude**.

3. Define **water table**.

4. How do the terms **porosity** and **permeability** relate to aquifers and aquicludes?

5. What is the relationship between the zone of aeration, the zone of saturation, and the water table (draw a sketch).

6. What factors contribute to a rising or falling water table?

7. What conditions are necessary to form a **free-flowing** vs. **artesian spring**?

8. Label the following components in Figure 10.2.

   A. Aquifer and aquiclude(s)

   B. Possible place for a natural spring

   C. Place and depth of an artesian spring or well

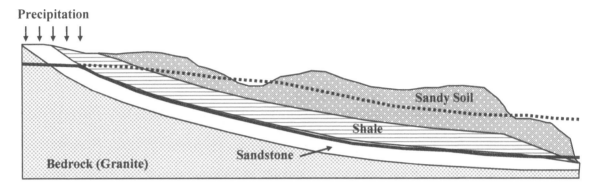

Figure 10.2

9. What is an artesian well and how does it operate?

10. Most residential homes and farms away from the city have their own well. What are some of the advantages and disadvantages of owning your own well?

Section III Hydrothermal Activity

1. What accounts for the hydrothermal activity in Yellowstone National Park?

2. Explain how a geyser operates.

3. How does a fumarole differ from a geyser?

Section IV Karst Landscapes

Water entering the ground is very ineffective in terms of hydraulic power. Instead these slowly moving waters chemically attack and dissolve certain rock units, like limestone. This solution process, called carbonation, results in landscapes known as **karst**, consisting of many solution pits, caverns, and underground drainage ways.

1. Explain the two-stage process of carbonation.

   A. $CO_2 + H_2O \Rightarrow H_2CO_3$
      (Carbon Dioxide + Water $\Rightarrow$ Carbonic Acid)

   B. $CaCO_3 + H_2 + CO_2 \Rightarrow Ca(HCO_3)_2$
      (Calcite + Carbonic Acid $\Rightarrow$ (Calcium Bicarbinate : a salt))

2. How do joints and bedding planes contribute to the successes of the solution process?

3. Name and define three features associated with karst visible at the surface.

   _____

   _____

   _____

4. Name and define three types of **dripstone**.

   _____

   _____

   _____

5. In what parts of the United States and world are Karst landscapes found.

Landforms Made by Streams Unit 11

Section I   Overland Flow and Streams

Precipitation that does not infiltrate into the ground will move as **overland flow** or as **stream flow**. The erosional energy of streamflow depends on the volume of water and the gradient of the streams which determine its velocity and turbulence. Once erosion takes place, the stream effectively transports its material load in one of three ways: dissolve load, suspended load, and bedload.

1. Describe each of these loads and give approximate percentages.

    A. _____

    B. _____

    C. _____

2. What happens to a stream's load as the velocity increases?

    A. As the velocity decreases?

3. Once deposition takes place, these stream-deposited material are referred to as

    _____.

4. Since the stream's flow varies with time it is important to know the stream's discharge at a given point in order to forecast flooding.

   A. Define discharge.

   B. How are peak discharges regulated to avoid floods.

Section II   Stream Drainage Patterns

1. On Figure 11.1 identify the following.

   A. Drainage Basin

   B. Drainage Divide

   C. Interfluve

2. Also label stream orders on Figure 11.1.

3. Stream patterns develop in response to the underlying geologic structure. Complete the statements below.

   A. Dendritic patterns develop when

   B. Trellis patterns develop when

   C. Radial patterns develop when

Figure 11.1

Section III Stream Valley Dynamics

1. Valley widening is primarily accomplished by the stream flow characteristic of

   _____

2. Valley lengthening is primarily accomplished by the stream flow characteristic of

   _____

3. An aerial view of a stream flowing in the direction of the arrow is given in Figure 11.2.

   A. Locate the point(s) where bank erosion is taking place.

   B. Locate the place(s) of stream deposition.

   C. Indicate by labeling the meander neck and the probable point of cutoff.

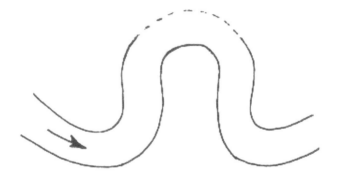

Figure 11.2

4. Label the following features of a stream valley landscape in Figure 11.3. Define the features listed below.

   A. Floodplain

   B. Meander

   C. Alluvial Deposits

   D. Undercut Slope

   E. Point Bar

   F. Oxbow Lake

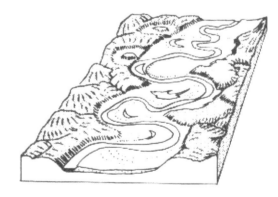

Figure 11.3

Section IV Stream Erosion

As mentioned in Section III, stream erosion affects all but the glacial landscapes on Earth's surface. Yet the landscapes sculptured by streams don't all look the same, because this sculpture depends on many geologic factors as well as time. It is convenient to divide the entire stream into three segments and look at their distinct landform types as shown in Figure 11.4.

A. Landforms in the upper reaches of the stream = upper course stream.

B. Landforms in the middle reaches of the stream = middle course stream.

C. Landforms in the lower reaches of the stream = lower course stream.

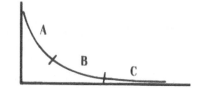

Figure 11.4

These typical landforms for the three segments of the stream can also be partially compared to W.M. Davis' concept of describing fluvial landscapes in terms of youthful, mature, and old.

1. List the characteristic landforms associated with **upper course or youthful streams**.

2. List the characteristics landforms associated with **middle course or mature streams**.

3. List the characteristics landforms associated with **lower course or old streams**.

Fluvial and Eolian Landforms of Deserts   Unit 12

Though some landforms have been sculpted into horizontally layered sedimentary rock creating plateaus, mesas, buttes or pinnacles, many others are associated with deformed and/or dipping sedimentary layers including cuestas, hogbacks, domes, and basins. All of these features are easily recognized in arid regions where the surface is exposed due to lack of vegetation and where running water has strongly sculpted the rock through differential erosion.

Section I   Fluvial Erosion on Horizontal Layered Rock

1. On Figure 12.1 identify the following items. Define the features listed below.

    A. Plateau

    B. Mesa

    C. Canyon

    D. Butte

    E. Cliff

    F. Badlands

    G. Bench

    Figure 12.1

2. Suggest the rock type which makes up:

    A. the top of the mesa _____

    B. the Badlands _____

3. Using Figure 12.2 Name the feature at:

    A. _____

    B. _____

    C. _____

    a. Horizontal Layers   **Figure 12.2**

Section II   Feature Identification

Identify the features in the photos below.

1.

2.

3.

Section III Stratification; Rock Layers, Erosion and Weathering

The cross section of stratified rocks over gneiss and schist depicted in Figure 12.3 is from the Grand Canyon. The rocks shown in the section include limestone, sandstone, shale, and schist.

1. Inspect the cross section. Arrange the rocks into two general classes: those you think might be relatively hard and most resistant to weathering and erosion and those you think might be relatively soft and least resistant. (**Note:** the distance between the faces of the rock layers exposed in the valley sides and the river is in part a function of time.)

    A. most resistant rock strata =

    B. least resistant rock strata =

2. What criteria did you use to differentiate between the most and least resistant rocks?

3. What generalization can you make from this example about the relationship between the characteristics of the rocks and the nature of the land surface?

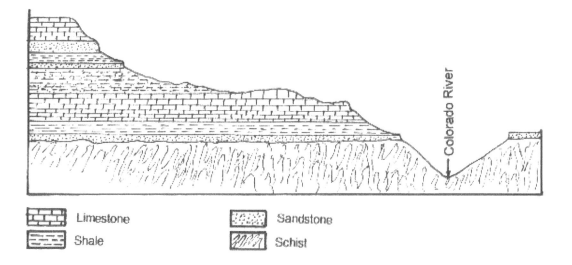

Figure 12.3

4. Earth's surface materials are continually affected by internal and external energy processes.

    A. List several internal processes.

    B. List several external processes.

Section IV Wind Erosion

1. In which terrestrial environments are wind erosion most active?

2. What is deflation and what landforms does it produce?

3. Define wind abrasion and give one example.

Section V Wind Deposition

1. Wind transports loose particles of small size and eventually deposits them into hills of loosely consolidated materials called dunes. Make a simple sketch of each type and show wind direction.

   A. Cresent (Barchan) Dune

   B. Transverse Dune

   C. Longitudinal Dune

2. Parabolic (blowout) dunes are common along Lake Michigan. Describe their shape and the manner in which they are formed.

3. Dunes are composed mostly of what type of materials?

Glacial Landscapes   Unit 13

Section I   Evidence of Glaciers

1. The Pleistocene started about _____ years ago and ended about _____ years ago.

2. At its greatest extent the ice sheet covered about _____ percent of Earth's surface.

3. Give three pieces of evidence for the former existence of continental glaciers.

   A. _____

   B. _____

   C. _____

4. What are some of the hypotheses which aim to explain the growth of continental glaciers?

   A. _____

   B. _____

   C. _____

5. What is meant by the term **glacial stage** during the Pleistocene Epoch?

6. What is meant by the term **glacial interval** during the Pleistocene Epoch?

7. Name the last glacial stage. _____

   A. How many years ago did the ice of that last stage reach its greatest extent? _____

   B. When did the ice finally retreat to its present position? _____

Section II   Mountain Glaciers

The effects of increased elevation are generally the cause for cooler temperatures and increased precipitation. These conditions are favorable for the formation of mountain glaciers. In many of the world's high mountain areas, glaciers have greatly altered the topography, creating some of our most beautiful and spectacular scenery. In the United States, mountain glaciation has been, and still is, active in the Pacific Northwest, and Alaska.

1. Indicate in Figure 13.1 where the following landform features are located. Define the features listed below.

   A. Cirque

   B. Cirque Lake or Tarn

   C. Arête

   D. Col

   E. Horn

   F. Glacial Trough

   G. Hanging Valley

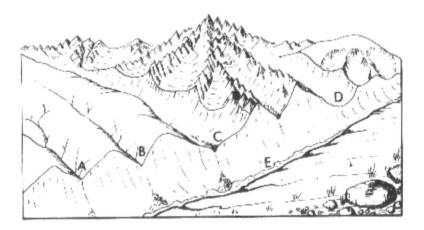

Figure 13.1

2. Note the streams marked A, B, C, D and E on Figure 13.1.

   A. Which valleys were primarily eroded by streams? _____

   B. Which valleys were eroded by glaciers? _____

3. Name the feature(s) in the pictures below.

A. _____

B. _____

C. _____

D. _____

Section III   Continental Glaciers

1. Correctly label each of these glacial landscape features in Figure 13.2. Define the features listed below.

    A. Terminal Moraine

    B. Recessional Moraine

    C. Interlobate Moraine

    D. Ground Moraine/Till Plain

    E. Outwash Plain

    F. Drumlin

    G. Esker

    H. Kettle

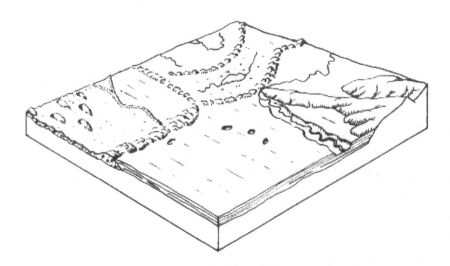

Figure 13.2

Section IV   Review

1. List the features created by stream erosion.

2. List the features created by stream deposition.

3. List the features created by wind erosion.

4. List the features created by wind deposition.

5. List the features created by continental glacial erosion.

6. List the features created by continental glacial deposition.

7. List the features created by alpine/mountain glacial erosion.

8. List the features created by alpine/mountain glacial deposition.

Appendix 1
Table of Physical Equivalents

This table is provided as a reference to the mathematical equivalence between many common physical measurement units. Please note that equivalence can be expressed in many ways, for instance (using a representative fraction as a map scale expression):

1:24,000	Correct expression of R.F. as a ratio
$\frac{1}{24,000}$	R.F. is literally a fraction (invertable)
1=24,000	Interpreting the ratio or fraction
One inch on the map equals 24,000 inches of actual ground distance.	English statement of the equivalence

You may be required to calculate values by multiplying correctly with any of these equivalents.

Linear Distance

1 Statute Mile (mi) = 1.609 Kilometers (km)
1 Statute Mile (mi) = 5,280 Feet (ft)
1 Statute Mile (mi) = 63,360 Inches (in)
1 Yard (yd) = 0.914 Meters (m)
1 Foot (ft) = 0.3048 Meters (m)
1 Inch (in) = 2.54 Centimeters (cm)

Area

1 Square Inch (in2)= 6.4516 Square Centimeters (cm2)
1 Square Foot (ft2) = 0.092 Square Meters (m2)
1 Square Yard (yd2) = 0.8361 Square Meters (m2)
1 Acre (a) = 0.4047 Hectares (ha)
1 Acre (a) = 4,047 Square Meters (m2)
1 Square Mile (mi2) = 2.59 Square Kilometers (km2)

Weight

1 Pound (lb) = 0.4536 Kilograms (kg)
1 Ounce (oz) = 28.35 Grams (g)
1 Ton (ton) = 0.907 Metric Tons (t)

Volume

1 Fluid Quart (qt) = 0.9463 Liters (l)
1 Dry Quart (qt) = 1.101 Liters (l)
1 U.S. Gallon (gal) = 3.785 Liters (l)
1 Cubic Yard (yd3) = 0.7646 Cubic Meters (m3)
1 Cubic Inch (in3) = 16.387 Cubic Centimeters (cm3 or CC)

Angular (note that this is also related to time measurement with hours, minutes, seconds)

1 Degree (o) = 60 Minutes (') = 3600 Seconds (")
1 Minute (') = 60 Seconds (")

Temperature

0 C = 5/9 (0 F - 32)
0 F = (9/5 0 C) + 32

Absolute Temperature 1 Kelvin = 0 C + 273.15   (not noted as 0 K; simply Kelvin or K)